WONDERS OF THE WORLD

EVERGLADES

Eileen Lucas

Technical Consultant
Fred E. Dayhoff
Park Ranger
Everglades National Park

Primary Photographer
Cheryl Koenig Morgan

RSVP

RAINTREE
STECK-VAUGHN
P U B L I S H E R S
The Steck-Vaughn Company

Austin, Texas

A production of B&B Publishing, Inc.

Editor – Jean B. Black
Photo Editor – Margie Benson
Computer Specialist – Katy O'Shea
Interior Design – Scott Davis

Raintree Steck-Vaughn Publishing Staff

Project Editor – Helene Resky
Project Manager – Joyce Spicer

LIBRARY OF CONGRESS CATALOGING-IN-PUBLICATION DATA

Lucas, Eileen.
    Everglades / by Eileen Lucas
        p. cm. — (Wonders of the world)
    Includes bibliographical references and index.
    ISBN 0-8114-6373-7
    1. Natural history — Florida — Everglades — Juvenile literature.
2. Wetland conservation — Florida — Everglades — Juvenile literature.
3. Everglades (Fla.) — Juvenile literature. [1. Natural History — Florida —
Everglades. 2. Everglades (Fla.) 3. Conservation of natural resources.]
I. Title. II. Series.
QH105.F6L83 1995
508. 759'39 — dc20

94-26278
CIP
AC

Cover photo
Great blue heron at Anhinga Trail

Title page photo
A scenic view of the Florida Everglades

Table of Contents page photo
An alligator in the Everglades

PHOTO SOURCES
Cover Photo: © Cheryl Koenig Morgan
© Tom Boyden: 26 top, 34 top, 35
© John Domont: 34 bottom
Florida Department of Commerce/Division of Tourism: 8 bottom, 58, 60 right
Florida Marine Research Institute: 60 left
Florida State Archives: 37, 38, 39 right, 41
Courtesy Homosassa Springs State Wildife Park: 59
© Gary Kramer: 15, 24 both, 29, 31, 32 both, 53 inset
© Gene Li/South Florida Water Management District: 1, 6 left, 47
Eileen Lucas: 10 inset

© 1987 Cheryl Koenig Morgan: 8 top, 10 bottom, 11, 12, 13, 14 both, 16 both, 17 both, 18 both, 19, 20, 22 both, 25, 26 bottom, 27, 28 both, 30, 33, 36 both, 39 bottom, 43, 46 top
National Museum of American Art, Smithsonian Institution: 40
© Pat Parrington/South Florida Water Management District: 6 top, 23, 53 bottom
© Cindy Pelescak/South Florida Water Management District: 3, 48 top
South Florida Water Management District: 21, 42, 44, 45 both, 46 bottom, 49, 50 both, 51, 52, 55, 56, 57, 61
© Alan Briere: 5

Printed and bound in the United States of America.
1 2 3 4 5 6 7 8 9 10 VH 99 98 97 96 95

Table of Contents

A Very Special Place

Imagine for a moment that you are in a canoe, paddling down a narrow channel of water that opens before you. You are moving through great vistas of grass that spread out as far as the eye can see. Then the sea of grass is broken by a small island. Its gnarled trees and hanging vines seem to rise magically out of the water. Pulling your canoe to the bank, you step carefully ashore and begin to walk quietly among the trees.

It is a hot summer day. Mosquitoes buzz all around, and birds cry out in the trees, warning unseen creatures of your approach. Coming upon a small pool among the dark trees, you pause at the sight of a tall white bird standing motionless in the water. Oblivious to your approach, it suddenly bends its head and, with one swift thrust, catches a fish. It doesn't seem to notice what looks like a log floating its way. Your eyes widen as you realize that the "log" is actually an alligator. Luckily, something tells the bird it's time to leave, and its widespread wings lift it away from the pool and the hungry alligator. The large reptile floats on, preparing to catch some other animal unaware, perhaps a turtle or even a raccoon. The graceful bird—a great white heron—disappears from sight.

It is hard to believe that the beautiful, primitive scene around you is not some distant, exotic land. Instead, this subtropical world lies just a few hours from the huge metropolitan area of Miami, Florida, in a very special place called the Everglades.

A Place of Water

The Everglades, often called the Glades, is a wetland—a large region in southern Florida where the low-lying, flat land is saturated with water all or much of the year. This wetland starts in the Kissimmee River Valley, just above Lake Okeechobee. The water then spreads out to both the east and the west coasts of the

"Leaving the trail I worked my way out into the dense, tangled growth, and as I sat down at the foot of a great tree and gazed around and upward it seemed to me as though the spirit of the forest took possession of me. . . . I sat and fairly drank in the wonderful silence and loneliness of the hammock. In such a place one must be alone to enjoy the full beauty and sweetness of it all."

— Charles Torrey Simpson, describing a visit to a hardwood hammock, quoted in *Some Kind of Paradise*, by Mark Derr

Georgia

Atlantic
Ocean

Gulf of Mexico

Florida

Lake
Okeechobee

**Everglades
National Park**

The Everglades is a land of water, grasses, and sky. This is Taylor Slough, a low-lying region of slowly moving water.

Tree islands called hammocks are found in parts of the Everglades where the land rises above the water (right). Visitors to Everglades National Park can walk across boardwalks built over saw grass to get to the hammocks (*below*).

peninsula where it moves southward to Florida Bay.

The wetland making up the Everglades consists of marsh and swamp. A marsh is an unforested wetland covered largely with grasses and very few woody trees, while a swamp is forested wetland with enough dry spots for trees and other plants to grow.

Although the Everglades has been called the River of Grass, it is unlike any other river. At about 120 miles (193 km) in length, it is not especially long. It is, however, exceptionally wide, averaging between 40 and 60 miles (64 to 97 km) across. And it is very shallow, averaging only about 6 inches (15.2 cm) deep, though water levels rise and fall considerably.

The water of the Everglades does, indeed, flow like a river, but it moves so slowly that you probably wouldn't realize it is flowing at all. It creeps southward at a rate of no more than 1,320 feet (402 m) per day. However, during high-water periods, in the open deep sloughs where there isn't much vegetation to slow it, the water can move many thousands of feet each day.

Wetlands are important for several reasons. They provide a special place where a multitude of plants and animals can live, and where migrating birds can stop for food and rest. In addition, wetlands control the water in a region. They help send water back into the underground water supplies, and they filter out pollutants as water runs into rivers and lakes or seeps underground.

There are, of course, other wetland areas in the United States, but this one is different. The climate of southern Florida is similar enough to the tropical regions near the equator that some tropical plants and animals can live here. In some parts of the

Everglades, tropical plants and animals live side by side with plants and animals that are more typically found in the north. This combination creates a wildlife region that is unique—there is no other place on Earth quite like it.

For a long time, people have recognized that this unique environment must be protected. Part of the area was designated a national park in 1947. Everglades National Park covers 2,746 square miles (7,112 sq km), making it one of the largest national parks in the lower 48 states.

The remainder of the Everglades and the adjoining swamplands have been divided up in many ways. Some of it belongs to Water Conservation Districts, which are like giant holding ponds for water. Some has become part of the Big Cypress National Preserve. Some has been set aside as reservations for Native Americans. And some is owned by conservation groups, such as the National Audubon Society and The Nature Conservancy.

Unfortunately, about one-half of the original Everglades has been drained to create dry land for

Southern Florida and the Everglades Region

construction and farming. These areas no longer look or act like wetlands. Many individuals and organizations have now joined forces to protect the remaining Everglades wetlands.

Condominiums line a canal in an exclusive southern Florida neighborhood. Wetlands were drained to make the land "useful."

The Origin of the Everglades

Millions of years ago, when glaciers covered the northern part of the North American continent, the Atlantic Ocean covered all of Florida. About 19 million years ago, the Earth entered a very cold period, and the glaciers started to expand and move southward. As more and more moisture became trapped in the glaciers, the water level of the ocean dropped, and a peninsula of land, called the Florida Plateau, was exposed at the southeastern tip of the continent.

This is the Florida peninsula as seen from space. The large dark patch in the center is Lake Okeechobee.

Many different animals sought refuge from the glaciers here among the tropical trees and plants.

Slowly, as the Earth warmed, and the glaciers began to shrink, melting ice caused the water level of the ocean to rise once more, covering the Florida Plateau. Over hundreds of thousands of years, this process was repeated numerous times. Each time the water covered the plateau, the animal life in the ocean left behind thick layers of shells made of the mineral calcium carbonate, which compressed to form the rock called limestone.

When the waters retreated for the last time, about 17,000 years ago, the surface of the land was relatively flat except for a slight rise along the eastern coast. The land sloped gradually southward, only a few inches per mile, somewhat like a slightly tipped saucer, until it dropped below the surface of the ocean at the southern tip.

The foundation layer of limestone was pitted and porous, something like the surface of the moon. On top of this, layers of decaying plant and animal matter were deposited, forming a spongy substance called peat. Grasses and other plants grew all across the land. As thick mats of muck accumulated in the grass roots, areas of slightly higher ground were created where trees could grow. A multitude of plants and animals began to prosper in this semitropical place.

Water, Water Everywhere

At the northern end of what would become the Everglades, a large, shallow lake formed. The Native Americans called it "Okeechobee," meaning "Big Water." Shaped like a huge soup bowl with over 135 miles (217 km) of shoreline, Lake Okeechobee was only 14 to 15 feet (4.3 to 4.6 m) deep.

When the rains came, Okeechobee often overflowed. Then a huge sheet of water ran southward from its banks into the grassy lowlands. With an annual rainfall near 60 inches (152 cm) and the overflow from Lake Okeechobee, most of the land was wet for most of the year. In a few places, slight troughs, or grooves, formed on the surface of the land, directing the water flow. Some of the water went southeastward, into the Atlantic Ocean along Biscayne Bay. Some headed straight south into the blue-green waters of Florida Bay. And much of it moved southwestward, finding its way to the Gulf of Mexico.

Over the centuries, some of the water dribbled and percolated down through the soil. It continued down through holes in the porous limestone rock, creating an aquifer, a sort of underground lake or reservoir. This reservoir is called the Biscayne Aquifer.

However, the thick layers of peat and marl (a mixture of clay, minerals, and shells) that had formed over the limestone kept much of the moisture on the surface of the land. Most of this water evaporated into the air, where it gathered in the atmosphere and fell again as rain, perpetuating the cycles of wetness.

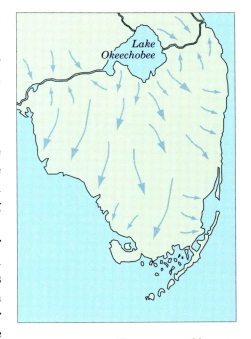

The arrows on this map show the historic water flow of southern Florida. Beginning at Lake Okeechobee, water would overflow and run to the south, southeast, and southwest.

This is one of the most important functions of the Everglades—providing a place where water can both trickle down to the aquifer and evaporate up into the air. Without the Everglades, there would be less rainfall over south-central Florida—and much less fresh water for its people.

Seasons and Weather

There are two seasons in south Florida: rainy and dry. During the dry season—roughly from November to May—there is very little rainfall. Temperatures generally range from a cool 50°F to a warmer 70°F (10° to 21°C).

The wet season, which lasts from approximately June to November, is marked by heavy rainfall and high temperatures, usually 80° to 90°F (27° to 32°C). About 50 to 60 inches (127 to 152 cm) of rain fall over south Florida during this season. As Everglades writer Archie Carr says in his book *The Everglades*, "Rain is life itself in the Glades, and no creature there minds getting a little wet." The rains cause the Kissimmee River to swell and Lake Okeechobee to overflow and flood the River of Grass year after year. It is this fluctuation of wet and dry, this natural balance, that defines the Everglades.

During the rainy season—especially from late August to early October—hurricanes may hit the Glades. A hurricane is a powerful, whirling storm of wind and water that forms in tropical areas of the

Thick water lettuce grows in "lakes" during the rainy season *(inset)*. As the lettuce dies, it is deposited on the bottom of these lakes and becomes part of the rich organic peat. In the bottom picture, an alligator trail winds through the dark peat of a dry lettuce lake.

A HURRICANE NAMED ANDREW

On August 24, 1992, a hurricane named Andrew devastated a large portion of south Florida. It struck just south of Miami, ripped through the town of Homestead, tore across Everglades National Park, and left Florida just south of Naples to move across the Gulf of Mexico to Louisiana, where it struck again.

Along that path, more than 80,000 houses and other buildings were destroyed or badly damaged, with $20 billion in property damage. Wind gusts were estimated at nearly 200 miles per hour (322 kph)—they had to be estimated because measuring devices were knocked out by the storm. At the time, Andrew was called "the most devastating natural disaster ever to strike the United States." Miraculously, only 65 lives were lost (43 in Florida) because most of the people in the storm's path heeded warnings to seek safety.

For the plants and animals of the Everglades, however, there was no escape. Many of the hardwood trees in Everglades National Park were uprooted or damaged. At least 70,000 acres (28,328 ha) of shoreline trees called mangroves were destroyed.

Under natural conditions, the Everglades should be able to recover from such damage. But the region's complicated ecosystem has been so altered by human activities that naturalists and other scientists are concerned. People who care about the Everglades will be monitoring its recovery very carefully.

oceans when warm, moist air rises quickly and cooler air rushes in below it. Rapidly swirling winds are created, accompanied by extremely heavy rainfall. While the storm itself travels at only about 10 to 30 mph (16.1 to 48 kph), its wind speeds are usually well over 100 mph (161 kph) and can sometimes be up to twice that speed.

Hurricanes have struck both coasts of Florida and moved across the Everglades many times. Although they cause extensive damage to plants and disrupt the habitats of animals, they are natural phenomena—a normal part of the cycle of life and death. Some of the plants of the Everglades—and perhaps even a few of the animals—may have arrived there on the winds and waters of hurricanes.

Writing in *Everglades; The Story Behind the Scenery*, author Jack de Golia notes, "The day after a hurricane, when havoc is everywhere, it seems the storm was nothing but an exercise in brute destruction. Viewed from the perspective of centuries of such activity, however, it is evident that hurricanes actually provide opportunities for living things to increase—by opening up new places for sun-loving plants; by churning nutrients from marine bottoms to make them available to the pink shrimp, fish, crabs, and lobsters in the waters above; and by carrying plants and animals to new habitations."

Chapter Two

Worlds Within

The special world of the Everglades is actually made up of several different regions. Each of these regions can be looked at as a small ecosystem within the total Everglades ecosystem. An ecosystem is an interacting combination of plants and animals and the physical environment they inhabit.

The central low-lying area of the Everglades—the true Glades—is made up of several bodies of water called sloughs. These are shallow indentations in the land where water drains and gathers.

The main tributary of this water system is known as Shark River Slough. The water in this slough flows gradually southwestward into the Gulf of Mexico. Lostman's Slough, the next largest tributary, supplies water to the entire southwest corner of Everglades National Park. Most of Lostman's Slough actually lies outside of the national park in Big Cypress Preserve. A smaller tributary—Taylor Slough—drains southeastward. Scattered throughout this marshy area are many small islands of trees.

The Shark River Slough runs through the center of the Everglades, providing fresh water to the grasses on each side.

Along the eastern coast is a raised ridge of land that became a pine forest. To the northwest lies the Big Cypress wetland region, a part of the Everglades ecosystem that is separated from the central low-lying area (and from Everglades National Park). Along the southern and western coasts of the Everglades is an area of swamps where saltwater-loving trees called mangroves grow. The northwestern part of this area is also known as Ten Thousand Islands because it consists of many small mangrove-covered islands. At Florida Bay, the extreme southern portion of the Everglades, the delicate ecosystem of the Everglades meets the saltwater ocean.

All these parts of the Everglades are entwined in an interconnected ecosystem. It is an ecosystem in which the form of the land and its living things are closely interwoven. Rock formations determine water levels, which in turn determine what sorts of plant and animal life will live there.

The True Glades

The true Everglades—the River of Grass—is a flat stretch of land that is thickly covered with grass and sedge, a grasslike plant with a solid shaft, rather than the hollow shaft common to grasses, and dotted with islands of trees. This is the land the Native Americans called "Pahayokee," or "Grassy Waters."

Unlike true grasses, saw grass stems are solid. Sharp, toothlike projections line the edges of the blades.

These miles upon miles of marsh are home to some of the world's most beautiful and interesting plants and animals.

Approximately 70 percent of the plants in the marshes and sloughs of the true Everglades is saw grass, the plant that gives the Everglades its nickname—the River of Grass. Saw grass, a type of sedge, gets its name from the rows of tiny, sharp teeth that grow on each edge and down its midrib. Without the proper protection, a person walking through saw grass can end up with tattered clothing and torn skin.

At one time some 3 million acres (1.2 million ha) of saw grass covered the Everglades. Along the shores of Lake Okeechobee, where the Everglades was traditionally at its wettest, the saw grass often grew to heights of 14 feet (4.3 m). As it died, collapsed, and decayed, it created a layer of fertile muck that enticed early settlers to farm it. Now known as the Everglades Agricultural Area, most of this land has been converted into sugarcane and vegetable farms.

Growing at the base of the saw grass stems is a small algaelike plant called periphyton. One of the tiniest Everglades plants, it has an extremely important job in the ecosystem.

Masses of periphyton cling to the base of saw grass stems and other plants of the Everglades. When the dry season comes and water levels drop, the periphyton slides down the blades of grass and settles on the floor of the marsh. Embedded in this mat of plant life are the seeds and eggs of many small plants and animals. While the marsh around it is drying out, the periphyton holds and protects enough life to repopulate the Everglades when the summer rains come again.

The saw grass marshes of the Everglades are rich in many forms of animal life. Insects, fish, amphibians, reptiles, and even mammals make their homes in this environment. The water rat builds a nest of dry grass along the edges of the sloughs, river otters play in the shallow water, and frogs fill the air with the sound of their calls.

Besides ensuring that there will be more frogs, frog eggs are an important source of food for many animals. That is why it is good that they lay so many—800 at a time for

Masses of tiny plants called periphyton attach themselves to the saw grass that grows in the fresh water of the Everglades.

the tiny tree frog, as many as 20,000 for the big bull frog. Dragonflies are among the creatures that eat these eggs. It's not a one-sided feast, however—adult frogs eat dragonflies.

One of the key residents of the Everglades is the alligator. As it crawls through the fields of saw grass, the alligator makes trails that can be followed by other animals. Pools of water, called gator holes, in which alligators spend the long hot days of summer, are sanctuaries for many other water-loving animals, too.

Hammocks and Heads

After saw grass, the most dominant plants of the Everglades are trees. More than 120 species of trees thrive in the Everglades. Some take the high spots, while others prefer low, wet areas. A difference of a few inches in elevation can lead to the development of dramatically different habitats.

Tropical hardwood trees, with such odd names as gumbo-limbo, poisonwood, pigeon plum, live oak, red bay, bustic, and mastic, grow on elevated spots of land rising perhaps 1 to 2 feet (30 to 61 cm) above the surface of the River of Grass. Hundreds of these tree islands, called hammocks, are scattered throughout the Everglades.

Like a moat protecting a castle, the water flowing around hardwood hammocks helps to protect the trees from fires that sweep across the saw grass marsh during the dry season. These "moats" are

Alligators live along the edges of waterways. These reptiles are usually between 6 and 12 feet (1.8 to 3.7 m) long, although some may reach 19 feet (5.8 m).

Tree islands—areas of higher ground—are called hammocks.

formed when acid from decaying plant matter eats away the surrounding limestone. Except during the harshest droughts, there is enough water in depressions in and around hammocks to protect the trees.

The hardwoods are mostly tropical trees that came from the West Indies in the Caribbean Sea. Their seeds were carried to Florida on the winds of storms or by ocean currents many years ago.

Some people think the gumbo-limbo tree looks as if it had sunburn because its reddish bark peels off in thin strips. Poisonwood is related to the poison ivy plant—as people who let its sap touch their skin have found out. A few mahogany trees can also be found, though many of the largest were cut down in earlier times to make fine furniture. One stand of living mahogany trees with trunks up to 4 feet (1.2 m) in diameter has been preserved in Everglades National Park.

Several types of palms can be found among the hardwoods on tree hammocks or growing by themselves in patches. The royal palm is a tall, stately, gray-white tree, topped by a green stem and long green fronds. The cabbage palm is less impressive, except of course to the animals—and people—who like to eat the "cabbage" that grows at the center of its fronds.

While the hardwoods have taken to the high and dry spots of the Everglades, cypress trees pre-

The unique gumbo-limbo tree has been nicknamed "tourist tree" because its red, peeling bark looks like a sunburned tourist.

fer the depressions. Even in the dry season, when the water level in the saw grass marsh drops, some shallow pools of water remain in parts of the Everglades, and this is where the cypress prospers.

Cypress trees often grow in relatively small, round stands called cypress heads. Cypresses also grow in longer, narrower stands called strands. A strand of cypress in Corkscrew Swamp Sanctuary is 3 miles (4.8 km) wide and at one time was over 20 miles (32.2 km) long.

The bald cypress is a deciduous tree, meaning that it loses its leaves in winter. The pond, or dwarf, cypress is much smaller—though often older—than the bald cypress. These two types of cypress may actually be the same species with different adaptations to particular soil and water conditions.

Around the base of many cypress trees, especially the larger ones, are knobby projections called knees. Some botanists believe that the cypress uses its "knees" to "breathe," since its roots are always submerged in water. Others think the knees support the tree. Both groups agree that the knees do not become new cypress trees, even though many other small plants may grow on and around these knees.

A host of interesting plants grows among the various kinds of trees in the Everglades. Pop ash and custard apple are two of the smaller trees that grow in the tangled forest. Several kinds of ferns are found in the shady hammocks, including the resurrection fern, which gets its name from its ability to go from a brown, dried-up looking mass to a vibrant green plant when it rains. Swamp lilies and water lettuce are among the many plants that grow on the surface of the water. Vines with beautiful flowers are strung like garlands among the tropical hardwoods of many hammocks.

The large base of a giant bald cypress tree has projections called knees on which lichens, ferns, and grasses grow.

Unlike the hardwood trees of the hammocks, cypress trees grow in the wetter, lower areas of the Everglades. The trees in the picture are dwarf cypress, a much smaller relative of the giant bald cypress.

Everglades pine forests usually have an understory of saw palmetto. This important Everglades' habitat is home to some of the region's endangered animals, such as the Florida panther.

The Pine Ridge

The Pine Ridge is a strip of limestone rising up to 20 feet (6.1 m) above sea level—about as high as land in southern Florida gets. The ridge runs from north of Fort Lauderdale to Florida City, and extends westward from the coast into Everglades National Park.

The ridge is named for the pines that thrived here before the cities were built. Housing developments and shopping malls now cover much of the limestone, and most of the pine forest is gone. A few pieces have been saved, however, including Long Pine Key in Everglades National Park—one of the largest areas of unspoiled pine forest left in south Florida.

The slash pines that grow on this limestone ridge survive in the very thin soil that accumulates in potholes in the rock. Since this soil remains dry in all but the wettest seasons, making their habitat the most likely to catch fire, the pine trees have developed a thick, fire-resistant bark. In fact, they are not only fire resistant, but they also depend on fire to kill off the hardwood seedlings that would

Controlled fires are started by park rangers to renew the pine-woods ecosystem.

otherwise compete for moisture and sunlight in their habitat. In the few areas where the pine forests remain, biologists have found that a certain amount of controlled fire is required to protect the pine forests from being invaded by hardwoods.

The saw palmetto is a low-growing form of palm tree found among the pines. It too has developed fire resistance to help it survive. Though the top of the plant may look dead after a fire, its thick stem and roots, tucked safely into cracks in the limestone bedrock, usually survive to grow again.

Coontie is another interesting plant of the pinelands. Over the centuries, various tribes of Native Americans discovered that they could make a kind of flour from its starchy root.

Native Americans often came to the Pine Ridge to hunt, especially during the wet season when land animals sought dry places. Deer are still common in what is left of the pine forests, as are gray fox and gray squirrel. Gray squirrels are not natives of the Everglades, but they are common in central Florida. They were introduced to the Everglades by people and have almost completely displaced the native fox squirrel.

Wildflowers bloom profusely along the Pine

PIGGYBACK PLANTS

An interesting kind of plant that grows abundantly in the Everglades is the epiphyte, or air plant. The epiphytes have literally taken to the trees to escape the shady, wet floor of the hardwood hammocks. They grow high on tall trees and other sturdy plants without hurting them or taking anything from them. Epiphytes get their moisture and nourishment right from the air with no need for roots and soil.

One group of epiphytes are the bromeliads *(right)*, including Spanish moss and the stiff-leaved and banded wild pine, which is related to the pineapple. Many bromeliads have long leaves that form a cup capable of holding water through much of the dry season. Besides supplying the plant itself with moisture, these cups of water help support the small animals that live in the hammocks, such as frogs, salamanders, and snakes. Even birds and raccoons

may look to bromeliads for a drink and a meal during the winter.

At one time there were many kinds of orchids—another kind of epiphyte—in the Everglades. Today, there are only a few species left because they were overharvested, and their hammock habitats were destroyed. Some orchids were saved in the 1960s when concerned individuals (mostly Boy Scouts) rescued them from trees about to be bulldozed and tied them to trees in protected areas.

The strangler fig is one of the most bizarre epiphytes. It is one of the only epiphytes that is harmful to its host. It starts to grow when a fig seed is dropped onto a hardwood tree by a passing bird. As the fig plant grows, it sends roots down and around the host tree, until it eventually suffocates the tree and proceeds to grow in its place.

Ridge, especially in summer when the rains come to the area. The pine woods have the most abundant varieties of flowering plants in the Everglades, including purple-flowered blazing stars and crimson morning glories. Many kinds of butterflies live among the wildflowers.

The atala caterpillar, a species found only in the Everglades, eats the green leaves of the coontie plant. Native Americans made flour from the root of this plant.

Big Cypress Swamp

Big Cypress Swamp, a large area of land southwest of Lake Okeechobee, has the mysterious, moss-garlanded landscape that many people mistakenly expect to see in the true Glades. Though it is part of the larger Everglades ecosystem, Big Cypress Swamp is separate from the River of Grass and from the Everglades National Park. Shaped like a huge basin, Big Cypress Swamp once had large areas of standing water all year round—a perfect habitat for cypress trees. However, areas to the north and west have been drained, and levees, or barriers, have been built to the northeast. These changes have reduced the amount of water flowing through Big Cypress Swamp. Now the swamp dries up almost every year.

The word *Big* in its name refers more to the amount of land it covers—roughly 2,400 square miles (6,216 sq km)—than to the size of the trees. About half of the land that makes up Big Cypress Swamp was set aside as Big Cypress National Preserve in 1974. Some of the rest remains swamp, some is farm and pastureland, and some has been drained and developed.

While the Big Cypress area has many of the same plants and animals as other parts of the Everglades, a few are especially well known. The most obvious of these are the cypress trees, which are among the world's longest-living trees. Some of the cypress trees in Big Cypress National Preserve were growing there before Christopher Columbus set sail from Spain! They are more than 600 years old. Unfortunately, in the 1940s, thousands of trainloads of cypress trees were taken out of this swamp and shipped north to build boats, bleachers, and pickle barrels. Many of those trees were over 100 feet (30 m) tall.

Along the southwestern edge of Big Cypress lies an area known as Fakahatchee Strand. The strand is 25 miles long and 7 miles wide (40 by 11.3 km). Though most of its biggest trees were cut and hauled out a half century ago, it remains a beautiful and mysterious wilderness area that is now protected by the state of Florida. The few black bears left in the Everglades region live in the strand.

Another part of Big Cypress that has been preserved is the Corkscrew Swamp Sanctuary. It holds the largest remaining stand of bald cypresses, some of which are over 700 years old. As in other parts of Big Cypress, there are also some lofty royal palms scattered among the bald cypress. Lush ferns grow on old cypress knees, and orchids and other epiphytes cover the tree trunks. Corkscrew Swamp is

A white-tailed deer stands in a field of saw grass at Corkscrew Swamp in Big Cypress National Preserve.

maintained by the National Audubon Society.

The Big Cypress Swamp area is important not only for the age of its trees but also as the preferred home of several very interesting endangered animals. These include the Florida panther and the wood stork. The few remaining panthers, a variety of puma or cougar, prey primarily on deer and small mammals. Wood storks catch fish in the shallow water beneath the cypress trees. These rare birds now nest only in the Corkscrew Swamp area of Big Cypress. Both of these magnificent species have become endangered by the changes people have made in their habitat.

Mangrove Swamps

Running southward along the Gulf Coast from Big Cypress to Florida Bay is the land of the mangrove swamp. The mangrove trees are hardy salt-resistant trees that have adapted well to their watery habitat. This environment, where the fresh water of the Everglades meets and mixes with the salt water of the ocean, is called an estuary. Estuaries provide a breeding ground for many kinds of plants and animals that require just the right balance of fresh and salt water.

The estuaries of the Everglades are home to three kinds of mangrove trees—red, black, and white. The white mangrove grows farthest inland where the water is still mostly fresh and generally beyond the reach of the tides. A little nearer the sea are the black mangroves, which can more readily tolerate the salty waters that engulf them at high tide. Black mangroves are distinctive for the root projections, or pneumatophores, that surround them like patches of wild asparagus.

On the shore, and often in the sea itself, grow the red mangroves. Like someone standing on tiptoe, the red mangrove raises itself up on arching, stiltlike roots. These roots are a miniature ecosystem all their own, attracting plant matter and debris, called detritus, as well as many kinds of

Wood storks still nest in the Everglades, but some biologists think by the year 2000 this will no longer be possible. Since the natural wetland cycles have been changed by humans, there may not be enough fish for the endangered wood storks to eat during breeding season.

Red mangrove trees send out prop roots into the surrounding water. These roots provide nutrients for animals such as oysters that attach themselves to the roots.

animals. As detritus collects in the mangrove roots, new land is formed, and then mangrove seedlings sail forth like pioneers, extending the reach of land farther into the water.

A mature red mangrove produces cigar-shaped seedlings about 6 to 12 inches (15.2 to 30.5 cm) long. When these seedlings fall from the tree, a few take root nearby, increasing the density of the mangrove forest. Others float away to find new land. Mangrove seedlings may travel hundreds of miles before finding a place to put down roots.

Fallen mangrove leaves collect around the roots and help stabilize the growing tree. As the leaves decompose, they provide nourishment for tiny forms of life, which in turn become food for larger organisms, including oysters and shrimp. Raccoons and other small mammals hunt for this food among the roots. Insects buzz around the leaves, occasionally serving as food for frogs. Meanwhile, the upper branches of the mangrove provide a home for many birds. Brown pelicans, for example, build their nests primarily in mangroves.

Florida Bay

The final Everglades ecosystem is Florida Bay—the area off the southern tip of Florida between the Everglades and the Florida Keys.

The mangrove swamps of Florida Bay provide habitat for sea animals such as lobsters and shrimp. In this region, fresh water flowing from the land and salt water from the sea mix together.

The roseate spoonbill is one of many beautiful wading birds in the Everglades. During the dry season from November to May, these birds feast on fish and crustaceans concentrated in the shallow water.

Many kinds of grasses and seaweeds grow in its warm, shallow waters. Florida Bay is also the site of North America's only living coral reef. The reef lies in John Pennekamp Coral Reef State Park, an enchanting underwater area. The grasses, seaweeds, and coral provide food and shelter for many fish, such as tarpon and snapper, as well as such crustaceans as shrimp and spiny lobster.

The beaches of Florida Bay are a breeding ground for ocean-dwelling loggerhead turtles and crocodiles. Bottle-nosed dolphins and endangered manatees swim here as well as along both coasts. Roseate spoonbills and great white herons fly over the bay as they return to their nests after a day spent fishing.

All these animals depend on the protection Florida Bay receives as part of Everglades National Park. But Florida Bay was not included in the original boundaries when the park was dedicated in 1947. It was added in the 1950s when its importance as a nesting and breeding area for so many kinds of wildlife was recognized.

In recent years, however, the decreased flow of fresh water through the Everglades and into the bay has created an ecological crisis. Plants that depend on a stable balance of fresh and salt water have begun to die off, reducing the habitat and food supply for aquatic animals. Water-management practices north of Everglades National Park are now being evaluated.

As many as 100,000 wading birds come to the Everglades every year to breed. Gulls live in the region all year.

Chapter Three

Everglades Wildlife

nimals come in all shapes and sizes in the Everglades. Some range across hundreds of acres and move through a variety of habitats, while others never leave a particular hammock or other small area.

Invertebrates

Invertebrates are animals that lack a spinal column, or backbone. Among the tiniest—yet most important—invertebrates of the Everglades are insects and their larvae. While swarms of mosquitoes make the Everglades close to unbearable for most people during the height of the rainy season, they are a critical element in the food chain. Their larvae provide food for small fish, and the adults are eaten by lizards, frogs, and bats.

Several hundred kinds of butterflies also live in south Florida, including some that are seldom found outside the Everglades and some that are in danger of becoming extinct. Butterflies are most abundant among the wildflowers of the pine forests. The greatest threat to these endangered species is their dwindling habitat.

Oysters, crabs, and shrimp are some of the aquatic invertebrates of the Everglades. These crustaceans, commonly found among the mangrove roots of the estuaries, are as important to the food chain as their land-bound counterparts.

Animals of the River and Swamp

It should not be surprising that in an environment as wet as the Everglades, fish are among the most numerous inhabitants. Tiny fish, a very important link in the food chain, feed on algae and insect larvae, while larger fish feed on smaller fish and are, in turn, eaten by birds, raccoons, turtles, and alligators.

Among the smallest Everglades fish is the tiny gambusia—also called the mosquito fish because it dines on mosquito larvae. Somewhat larger sports

The flocks of birds that covered the shelly beaches and those hovering overhead so astonished us that we could for awhile scarcely believe our eyes."

— John James Audubon, 1832, as quoted by William Robertson, Jr., in *Everglades: The Park Story*

A Halloween pennant dragonfly clings to a grass stem at Long Pine Key in Everglades National Park.

JEWELS OF THE EVERGLADES

One of the small but beautiful living things of the Everglades is the tree snail. Those of the genus *Liguus*, known for their beautiful colors, are found only in the hardwood hammocks, and some live only in specific hammocks. These snails originally came from the islands of Cuba and Hispaniola in the West Indies. Since moving to the mainland, however, the Florida tree snails have developed new colors and patterns on their shells. More than 50 different patterns, including stripes, splotches, and alternating colors, have been recorded.

Tree snails measure about 2 inches (5.1 cm) in length. They use their rough tongues to scrape lichens—primitive mosslike plants—from tropical hardwood trees. They seal themselves in their shells during the dry season to preserve moisture. During the rainy season, they move down the tree on which they live to mate and lay eggs.

Today, the beautiful tree snails are in danger of becoming extinct. Years of natural disasters, such as hurricanes and wild fires, and aerial spraying for mosquitoes, have significantly decreased the snail population. Destruction of marshland for housing and farms has led to loss of habitat for these small creatures. As a result, efforts have been made to transplant snails to safe hammocks inside Everglades National Park. Whether the snails will adjust and thrive in their new habitat is not yet known.

fish, such as bluegills and largemouth bass, attract fishermen to the Everglades. Even larger fish, including tarpon, snook, and barracuda, are found in the estuaries and Florida Bay.

Many fish die when the River of Grass dries up in late winter, but under normal conditions enough fish survive in watery places to repopulate the Everglades when it floods again in the summer. In recent years, however, human activities in the Everglades have endangered many aquatic animals. Their reduced numbers then endanger Everglades animals that depend on fish for food.

Other small animals of the river and swamp include reptiles and amphibians. Softshell and snapping turtles can be seen swimming in the waters of the Everglades or sunning themselves on fallen logs. Lizards are also common here, especially the Florida anole—a chameleon. Most of the lizards feed primarily on insects.

The Everglades is home to more than twenty varieties of snakes. Though most are harmless, the poisonous varieties are so fearsome that most visitors don't stray too far from designated trails. The diamondback rattlesnake, water moccasin, coral snake, and pygmy rattler are the venomous snakes. The indigo snake—a long, nonpoisonous snake

The endangered eastern indigo snake is not poisonous.

of the Everglades—is in danger of extinction.

Among the noisiest residents of the Everglades are the frogs. Each species has its own song to sing, and a rousing chorus can be heard on a rainy summer night. Says author Archie Carr, "Frogs do for the night what birds do for the day: they give it a voice."

Alligators—Keepers of the Glades

Alligators are probably the most notorious of the Everglades inhabitants—and rightly so. The alligator plays such an important role in the ecosystem that it is sometimes known as the "keeper of the Everglades."

Alligators are often seen sunning themselves on land to get warm, or resting in water to keep cool. These cold-blooded reptiles take on the temperature of their environment.

Alligators move through the water by using their powerful tails. At other times, they simply float on the surface of the water.

Alligators propel themselves through water by moving their powerful tail from side to side. Using its tail, snout, and legs, the alligator makes a den by clearing the plants from a small pool of water. It piles up the plant matter around the hole, creating dry ground on which other plants grow.

During the dry months of winter, when the water level in much of the Everglades drops, the alligator's hole becomes a refuge for animals seeking water. Without these gator holes, many Everglades animals would die during the dry season. The alligator will undoubtedly collect the rent by eating some of the tenants, but enough will survive to spread out once again in the Everglades when the rains come in summer.

When the female alligator decides it is time to lay her eggs, she uses her strong back legs and tail to push piles of plant matter into a nest that is usually about 4 feet wide by 6 feet long and 3 feet high (1.2 by 1.8 by 0.9 m). When the nest is ready, she climbs on top and lays about 30 to 40 eggs. She then covers the eggs with muck and leaves. The mother alligator tries to stay close to the nest, but she usually can't keep raccoons and other animals from eating some of the eggs.

When the baby alligators emerge from the eggs, they immediately head for the water of the gator hole. The mother alligator sometimes protects them with her presence while they are young. When she is not around, other adult alligators may eat the babies, particularly during low-water periods. Sometimes, adult alligators will investigate when a person imitates the distress call of a baby alligator. Many of the hatchlings also fall prey to large birds, otters, and other animals.

Alligators that survive their dangerous childhood grow quickly, gaining about 12 inches (30.5 cm) a year during their first few years. Adult males generally grow up to 13 feet (4 m), but some have been known to reach 19 feet (5.8 m).

A female alligator makes her nest of mud and plant material *(left)* during breeding season. She covers her hard-shelled eggs and guards them until the young alligators hatch *(below)*.

The American alligator is found in other parts of the southeastern United States beside the Everglades. Scientists estimate that there have been alligators in the area for 200 million years. There were probably about 5 million of them in Florida when the Spanish arrived and called the alligator *el legardo*—the lizard. However, people have been so efficient at hunting this lizard that in the early 1900s, it was thought to be in danger of extinction. Hundreds of thousands of alligators were killed for their skins, which were then fashioned into boots, belts, and purses.

In 1969, the alligator became protected by the new Endangered Species Act, and since then it has made a dramatic comeback. Though it is no longer endangered, its future is still somewhat precarious because many of the animals it eats have declined in number, due largely to habitat loss. Only if the places where alligators live and the food they eat are protected can the survival of this fearsome and fascinating creature be assured.

Animals of the Air

For many people, thoughts of the Everglades bring images of its beautiful birds. About 300 different species of birds live in Everglades National Park alone. The importance of the Everglades as a nesting site for so many kinds of beautiful waterbirds was a key factor in the establishment of Everglades National Park.

The large wading birds and shorebirds are among the most popular. The pink flamingo, commonly associated with the Everglades and south Florida, does not actually live here naturally at all. It merely visits sometimes from Caribbean islands. Flamingos occasionally seen in the wild in the far southern part of the Everglades have usually escaped from one of the theme parks or zoos that import them.

Another pink bird, however, is a resident of the Everglades. The roseate spoonbill—also known as the pink curlew—swings its broad bill from side to side in the water, screening small fish and insects from the muddy water.

Both white and brown pelicans are found in the Everglades. The white pelican is one of the world's largest waterbirds, with a wingspan of up to 10 feet

The endangered brown pelican breeds along the coastal swamps of the Everglades.

(3 m). It is a freshwater bird, whereas the somewhat smaller brown pelican prefers saltwater areas, especially among the mangroves. The brown pelican is noted for its dramatic habit of diving into the water to catch fish. It is also known to sit around waiting for handouts from people who have been out fishing.

Several kinds of herons live in the swamp and marsh. The great white heron is the largest American heron. However, other herons, including the great blue and little blue, also go through a white phase, which makes identification confusing. Other white waterbirds found in the Everglades include the snowy and common egrets and the white ibis.

The anhinga, also called the snakebird because it swims with only its snakelike neck and head sticking out of the water, is a beautiful fishing bird. When it spots a fish, the anhinga spears it with its beak, tosses it in the air, opens its beak wide, and swallows the fish headfirst. After a swim, the anhinga has to sit in a tree or on shore with its

This beautiful black crowned night heron *(right)* wades in shallow water looking for fish and frogs to eat.

When the anhinga is on land, its wings are usually spread so they can dry. One of the most popular trails in the Everglades, the Anhinga Trail, is named after this fascinating bird.

wings spread out to dry since it does not have the natural oil that keeps ducks and other swimming birds from getting their feathers soaked.

The wood stork is the only true stork found in North America. It has a blue-gray bald head and long blue-gray legs. Its feathers are white edged with black. One of the primary breeding areas for wood storks in the Everglades is the Corkscrew Swamp Sanctuary. The wood stork fishes by wading with its open beak in the shallow water. When it feels a fish, it closes its beak, within .025 seconds!

FOR THE LOVE OF BIRDS

Among the most famous visitors to frontier Florida was John James Audubon. In 1831 he journeyed down the coast of Florida in a government boat, observing and painting many of the beautiful waterbirds nesting there. He was working on a book that was to include paintings of all the birds of America.

Some fifty years later, American women began wearing hats decorated with the beautiful feathers of waterbirds. Some hats even displayed whole stuffed birds. Egrets (right), herons, and other birds were slaughtered by the thousands in the Everglades, especially during the mating season, when their plumes were at their prettiest. This meant that their untended eggs were eaten by predators, or the newly hatched nestlings were left to starve. By the early 1900s, the beautiful wading birds of the Southeast were nearly extinct.

People began to protest the killing. Some of these people formed Audubon Societies, named in honor of the great painter. Later, some of these societies united to form the National Audubon Society, which worked to protect all birds. Laws were passed to ban the sale of bird feathers and to protect nesting sites. When an Audubon warden named Guy Bradley was murdered by plume hunters in 1905, the tide turned against the bird killers. Plumed hats went out of style, and slowly, the population of wading birds began to recover. But the damage had been done, and they would never again be as numerous as they were in the days of John James Audubon.

In recent years, the most endangered bird in the area has been the Everglades kite, also called the snail kite because its only food is the apple snail. An Audubon Society census completed several years ago estimated that there were only about 30 of these hawks left. One of its few remaining breeding areas was the Kissimmee River Valley and the shores of Lake Okeechobee. As the habitats of the kites' food—the apple snail—disappeared, so did the Everglades kite. However, the Everglades kite has experienced a comeback, and hundreds now thrive in conservation areas including Shark River Slough and Lostman's Slough.

Scientists estimate that approximately 50 breeding pairs of American bald eagles currently nest in the Everglades. Interestingly, the bald eagles of Florida raise their young in winter, while eagles and other birds that nest in the north raise their young in summer. Bald eagles—endangered until recently—have been making a comeback all over North America.

Animals of the Woods

Many of the small land mammals found in the woods of North America can also be found in the Everglades. The opossum, which is common throughout the South and the Midwest, is one example. The raccoon is another. It occupies the role of middleman in the food chain, being both predator and prey to a number of kinds of animals.

Bald eagles nest in the Everglades.

In addition to the marsh rabbit, found throughout the Everglades, the northern portion of Big Cypress is also home to the cottontail rabbit. It can be found in upland areas and roadside glades. The marsh rabbit is darker in color than the cottontail, has shorter ears, and likes to swim!

Some people are surprised to learn that there are deer in the watery Everglades. The Florida variety of the white-tailed deer is somewhat smaller than those found farther north. Since white-tailed deer don't mind getting their feet wet, they are found in many areas of the Everglades. They often wade in belly-deep water, making their way from hammock to hammock. Higher water, however, will send them to dry spots—most notably the pine woods. In some parts of the Everglades, deer are so numerous that they can be legally hunted during a limited season, though not within the national park.

A few black bears live within the Everglades, but they are rarely ever seen. The best evidence of their presence is the occasional cabbage palm with its tasty "cabbage" bud torn out.

Another rare animal—now endangered because the Everglades is its only habitat left in North America—is the Florida panther. Between 30 and 50 of these animals are thought to be left in the Everglades, primarily in the Big Cypress area. The fur of the Florida panther is generally light brown or gray. Adult male panthers weigh as much as 150 pounds (68 kg), while females weigh somewhat

This radio-collared Florida panther is one of the few remaining in the Everglades. Scientists follow the movements of these animals very carefully, but the outlook for their survival is not good.

less. These beautiful animals like to roam across the cypress swamp, the hardwood hammocks, and the pine woods. Special underpasses have been built along some of the highways in their territory to protect them from cars.

Wildcats, relatives of the panther, are more common in the Everglades. They feed on rabbits, rats, and birds. The wildcat has long legs with large paws, a short body, and tufts of fur on its ears. These animals breed in the spring, bearing a litter of up to six kittens.

Manatees feed on the vegetation in the shallow water of rivers and along coastlines. Manatees are often killed by boat propellers as the watercraft traffic in Florida increases.

Ocean Animals

Many marine animals live in Florida Bay between the southern tip of the mainland and the Florida Keys. Because this area was included in the Everglades National Park, endangered sea creatures are protected here.

The loggerhead turtle is one such creature. Spending most of its life at sea, the loggerhead crawls up on the beaches of Cape Sable and other places along the southeastern Atlantic and Gulf coasts to lay its eggs each year. In the past, these nests were raided by humans and animals alike. Though people are no longer allowed to take these eggs, they still fall prey to raccoons and other predators. Also, like so many other animals, sea turtles lose their habitat when oceanfront property is developed or polluted.

Another endangered marine creature is the manatee, or sea cow. A strange-looking sea mammal, the manatee suffers greatly from the presence of people and boats in its water habitat. Because the manatee must swim near the water's surface in order to breathe, it is often struck by the propellers of small recreational boats. Also, it likes to feed on plants that grow in the shallow, warm waters of estuaries and canals where boat traffic is heaviest. Probably only a few hundred of these gentle, slow-moving animals

Dolphins are often found in the same territory as manatees, along Florida Bay and the Ten Thousand Islands. They can be seen jumping and diving in the wake of fast-moving boats.

remain in the waters off the coasts of the Everglades. However, the law now requires boats to go very slowly in manatee areas.

Another ocean animal occasionally seen on the beaches of Florida Bay is the American crocodile. Everglades National Park is the only place in the world where both alligators and crocodiles live. They are seldom seen together, however, as the crocodile never travels far from the salt water of the ocean, and alligators generally inhabit swamps, marshes, and lakes. The crocodile can be distinguished from the alligator by its long thin snout (the alligator's snout is more full and blunt), its gray-green color (the alligator is blackish), and its exposed teeth, which make it appear to be smiling.

Although the adult crocodile lives in the salt water of the ocean, it must nest near fresh water because newborn crocodiles require fresh or brackish (partly salty) water. Because the estuaries of Florida Bay were traditionally suited to their needs, 11,000 acres (4,452 ha) along the bay in the southeastern section of the park were set aside as a crocodile sanctuary in 1980. It is probably one of the best-protected crocodile habitats anywhere, and it may tip the balance in favor of survival of this ancient species in the Everglades. Several hundred of these reptiles are believed to live in and around the sanctuary.

Of course, the Everglades does not belong only to the animals. As we'll see in the next chapter, human beings have both lived in and exploited the River of Grass and its swamps.

A crocodile snout is narrower than an alligator snout. Also, its fourth tooth is exposed when its mouth is closed.

Chapter Four

Humans in the Everglades

The first humans came to the Florida peninsula well over 2,000 years ago. There was such an abundance of plant and animal life that some were able to settle in permanent villages without becoming farmers. Others harvested corn, beans, and squash.

The most prominent people to develop in the Everglades area were the Calusa. They used shells for tools and weapon tips with which they hunted deer, rabbits, alligators, and bears. They built open-sided homes with thatched palm roofs. They worshiped the sun and buried their dead in huge mounds of shells, sand, and dirt. They developed a rich and interesting culture in harmony with the Everglades for centuries.

Conquistadors

Then explorers set sail from the crowded ports of Europe in search of new routes to the East. Christopher Columbus landed on the islands southeast of Florida, and soon Spain controlled the resources of the Caribbean islands. Spanish soldiers

The Calusa used shells to make weapons and tools.

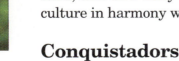

called conquistadors enslaved most of the native people, many of whom died from mistreatment. Many others died from European diseases, such as smallpox and influenza, against which they had no immunity.

News of the treachery on the islands must surely have reached the Calusa and their neighbors, for when Spanish ships landed along the coast of North America, the natives fought the invaders off. If the ships were wrecked in storms, the Calusa paddled out in their canoes, taking the gold and treasure, and killing or taking captive any survivors.

In 1513, a Spaniard named Juan Ponce de León left the Spanish-controlled island of Puerto Rico, looking for the land the natives called Bimini. Sailing northward past the Bahamas, he reached the North American coast (probably somewhere near Cape Canaveral) around Easter. Thinking he had discovered an island, he named the land "La Florida" either for the flowers he saw or because it was near Easter, called "Pascua Florida" in Spanish.

Looking for slaves and gold, Ponce de León moved southward along what is now the coast of Florida. When he met resistance from Native Americans, some of the Spaniards were wounded. After a few weeks, the explorers returned to Puerto Rico, not knowing that Florida was the southern tip of a huge continent.

Ponce de León returned to Florida in 1521, this time sailing along the west coast. As Spaniards swarmed from small boats onto what appeared to be a deserted beach, arrows suddenly shot from the forest. One of them wounded Ponce de León, and he died a short time later on the island of Cuba.

"I like to think of the Spaniards blazing their trails through the Florida hammocks. The hammocks were the same then as now, and will be the same forever if men can be induced to leave them alone."

— Marjorie Kinnan Rawlings in Cross Creek

Ponce de León discovered and named Florida. Some writers said he was searching for the fountain of youth.

For two hundred years, other Spanish explorers would claim but see very little of Florida. In 1763, Spain gave Florida to Great Britain in exchange for Cuba. By that time, the majority of the Native Americans living in south Florida—a region the Spanish had never really controlled—were dead. There was an unseen enemy the Native Americans had no way to fight. They died of European diseases by the hundreds, quietly, mysteriously, away from the eyes of historians. Perhaps a few Calusa lived on in secluded villages in the Everglades, and some may have fled to Caribbean islands. But many anthropologists believe they died out completely. We may never know for sure.

Groups of Seminole and Miccosukee still make their home in the Everglades today. They have found ways to blend traditional and modern activities into a single life-style.

A New Tribe

Any Calusa who may have survived eventually intermingled with the Seminole, a name which has come to mean any of the natives of Georgia and Alabama who moved into Florida to escape colonial rule. A majority of the Seminole were a Muskogee-speaking group of Creek people, called Cow Creeks, and their descendants. Others were Miccosukee, who spoke a completely different language and were sometimes at war with Muskogee-speaking Seminole. The Miccosukee tended to live farther south in Florida than the Seminole.

The Seminole were a people of many feasts, ceremonies, and rituals who followed family, clan, and tribal traditions. They generally lived in peace with the few Spanish citizens who stayed on in Florida as well as with the runaway slaves who reached their villages from the colonies to the north.

When the American colonies became the United States, they forced Great Britain to give Florida back to the Spanish. Relations between the new American government and the Seminole were strained. Southern planters wanted the slaves who had escaped to Florida returned to them. Many of these planters also wished to extend their land holdings into northern Florida. In 1817, the American government went to war against the Seminole, burning villages and taking slaves. The United States forces, led by General Andrew Jackson, killed many Seminole and Miccosukee and drove many others into the marshes of central Florida. This episode became known as the First Seminole War.

In 1821, the United States purchased all of Florida from Spain, and Andrew Jackson became the first American territorial governor of Florida. In 1830, as president of the United States, he called for removal of all Native Americans to land west of the Mississippi River and the return to slavery of all African Americans living with them.

Meanwhile, the Seminole and Miccosukee had adjusted to life in the marshes. Turtles, deer, and a flour made from a starchy root called coontie provided food. For shelter they built *chickees*—open-walled huts with raised floors and roofs made of palm leaves.

The Seminole and Miccosukee were now marsh-living people. In 1835, when the law required them to give up their

Seminole dugout canoes are made from cypress logs *(below)*. The front end is pointed so it can easily cut through the saw grass. The back end is flat so the boatman can stand *(right)* and pole the boat through shallow water.

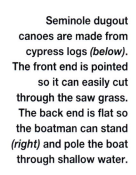

OSCEOLA — PATRIOT AND WARRIOR

When it became clear that the Seminole would have to fight for their land, a young Seminole stepped forward to lead his people. His name was Osceola, and he was a Muskogee Creek, born in Georgia in 1804. As a young boy, he moved with his mother to Florida, becoming part of the Seminole tribe. He married a woman who was descended from a runaway slave and a Native American. Though she had been born among the Seminole and had children with Osceola, she was captured at a trading post by slave catchers and taken into slavery in Georgia. This added to Osceola's growing distrust of the American government and his determination to fight for his people.

In 1832 Osceola was present when a U.S. agent was trying to force some Seminole leaders to sign a removal order. Osceola knew that the majority of his people did not want to move. He approached the table where the document sat and stabbed his knife in it, saying that it was the only way he would sign such a paper.

A short while later Osceola was arrested and jailed for opposing the removal order. It is said that he howled late into the night to show that he was not afraid of the soldiers. Upon his release, he began to prepare his fellow warriors for battle. In December 1835, he ambushed and killed the agent who had jailed him. That same day, another band of Seminole attacked a group of soldiers led by Major Francis Dade. The Second Seminole War had begun, and Osceola became one of its greatest leaders. He led his warriors into battle against American soldiers many times. However, he discouraged attacks against women and children, saying, "It is not upon them that we make war."

In 1837, while talking to soldiers under a flag of truce, Osceola was taken prisoner and was later sent to Fort Moultrie, South Carolina. While in prison there, his portrait was painted by the American artist George Catlin. Sick in body and spirit, Osceola died in prison on January 30, 1838. He was buried nearby. On his tombstone were written the words: PATRIOT AND WARRIOR.

lands and move to the arid plains of Oklahoma, they began preparations to fight for their land.

Led by a young warrior named Osceola, the Seminole attacked military posts and settlements. Under General Thomas Jessup, the U.S. Army attacked Seminole villages. These skirmishes became known as the Second Seminole War.

The war dragged on until 1842, after several thousand Native Americans had been removed to land west of the Mississippi. Those who remained, perhaps 300 to 500, lived on in the hidden hammocks of the Everglades. Some of these were small groups of Miccosukee. Others were Muskogee-speaking Seminole. Though they had often united in fighting against the soldiers, and their cultures were similar in many ways, they still thought of themselves as separate peoples. They continued to live in small isolated camps, finding food and shelter in the swamps and marshes.

Draining the Everglades

In 1845, Florida entered the United States as a state in which slaves could be owned. Slaveholders in the new state wanted more citizens to give them political power. However, new residents would need transportation, and most transportation methods required dry land. People began to talk about draining the Everglades.

A lawyer and scholar named Buckingham Smith was given the job of studying the Everglades. He concluded that for about $500,000 the region could be drained within a few years by cutting into the bedrock. His estimate would prove to be off by many years and millions of dollars.

The Swamp Lands Act of 1850 put much of the land of central and southern Florida under the control of the state. The law insisted that if the state was to control the land, it had to "reclaim" it.

The state of Florida sent armed men into the Big Cypress to survey and map the land. A Seminole named Billy Bowlegs resisted when soldiers entered his camp and vandalized his property. Once again Native Americans and settlers died, and villages were destroyed in fighting. This was the Third, and last, Seminole War. Eventually, many of the Seminole, including Billy Bowlegs, were forced to move west.

Still, a few Seminole and Miccosukee (between 100 and 300) managed to hold fast to their land deep in the Everglades, and, for the time being, they were left alone. They kept to themselves, venturing out of the Everglades on rare occasions to exchange feathers and animal skins for guns, cloth, and metal pots.

Meanwhile, throughout the 1880s, two businessmen—Henry B. Plant and Henry M. Flagler—bought up property and small railroads. They created great railroad and hotel empires, one down the west coast, and one down the east coast of Florida, ignoring the largely unknown interior of southern Florida, the land of the Everglades.

One person who set out to find the best way across the mysterious swampland in the 1890s wrote: "It may seem strange in our days of Arctic and African exploration for the general public to

Billy Bowlegs became a Seminole folk hero when he fought against U.S. Army soldiers that invaded Native American lands in Big Cypress. Many Seminole died; still others were forced to move west.

learn that in our midst we have a tract of land . . . that is as much unknown to the white man as the heart of Africa."

In 1905, Napoleon Bonaparte Broward ran for governor of Florida on a platform of "Drain the Everglades." Though some individuals called for further study, Broward replied, "I will be dead by that time. Let's get a few dredges and begin." He was elected, and the drainage began in earnest.

To drain away the water that created the River of Grass and the swamps, construction crews were instructed to dig canals through the land. Because water always seeks the lowest level possible, it flowed into the canals and gradually out to sea. The land along the canals dried and became "useful."

For a few years the weather was fairly dry, and people began to think that the Everglades were indeed being drained. Homes and farms were built. In the 1920s a highway called the Tamiami Trail was built across the Everglades, making it possible to travel overland from Miami to Tampa by way of Naples and Fort Myers.

To build the Tamiami Trail, men with machetes and axes walked through slough and swamp, cutting a trail. Others followed, using dynamite to reach the bedrock under the water and muck. The bedrock was then built up with loads of rock, and slowly, the road extended across the Everglades. Some workers lost their lives to explosions, drownings, and snakebites.

Dredging and draining the Everglades converted many acres of wetland into an agricultural paradise. Florida was soon called the "Vegetable Kingdom."

As people saw the Everglades apparently being tamed, they moved in and settled. The cities along the coast boomed, and the population of south Florida grew. As it always had when heavy rains came, water flowed down the Kissimmee River and Fisheating Creek and rose in Lake Okeechobee, spilling out into the land that had always been Everglades south of the lake. But now the land was dotted with farms, homes, and towns. The new residents, shocked and angry, complained about floods.

Then, on September 17, 1926, a hurricane came smashing down on Miami and ripped across south-central Florida. Thousands of homes were destroyed

and hundreds of people were killed. Two years later, on September 19, 1928, another deadly hurricane struck, one of the worst in Florida's history. Countless homes and 2,000 lives were lost in the wall of water that rose from Lake Okeechobee and drowned everything in its path.

It was obvious that the dredging had not been enough to stop the floods. A huge dike was built along the southern edge of Lake Okeechobee, sealing off its southern shore. The mound, they were sure, would keep the water in the lake.

A canal was cut westward to the Caloosahatchee River. Another was expanded on the St. Lucie River to the east. Now boats could travel across the state from Stuart on the east coast to Fort Myers on the west. And more importantly to the people worried about floods, the water level of Lake Okeechobee could be swiftly lowered by opening floodgates in the canals.

The water level of Lake Okeechobee dropped immediately when the gates to the canals were opened. The water bled off the land out to the Atlantic Ocean to the east and the Gulf of Mexico to the west. Thanks to the canals and the dike, Okeechobee would no longer overflow each rainy season. Water stopped flowing southward into the River of Grass, and it became a field of dried muck and sunburned grass. Farmers, ranchers, and city dwellers now had more dry land on which to live and work. They thought their problems were over. Instead, their problems with the Everglades were just beginning.

Today, huge southern Florida truck farms provide winter vegetables for the United States. They also use fresh water that once flowed freely through the Everglades.

Chapter Five

Trouble in Paradise

What people did not realize was that nature had many uses for the water that had flowed across the Everglades. One of them was the slow filtering down of water through the soil and peat to the limestone bedrock. This water accumulated in spaces in the limestone. These underground reservoirs, or aquifers, are the source of water drawn from wells. The Biscayne Aquifer under the Everglades provides fresh water for the people of south Florida.

As more and more Everglades land was drained, and as more and more wells were dug, there was less water coming into the aquifer. This meant that salt water from the ocean could now flow underground into the aquifer. Suddenly the water coming from many wells along the coast began to taste salty.

Salt water also got into water supplies through canals that had been dug into the limestone as part of the draining process. Ordinarily, fresh water flowed out of these canals to the sea. But when water levels in the canals were low, salt water could flow in.

By the 1980s, people began to realize that tampering with the water cycles of the Everglades was causing big problems. Not only was there too little water flowing through the Glades, but the water that did was polluted with chemicals from farm fields in the Agricultural Area.

Historically, fire has been a restorer of the Everglades ecosystem. When humans began tampering with the water flow, the Glades became very dry, and fires burned out of control.

In addition to sinking water levels, there was a problem with sinking land. As the natural moisture left the newly drained muck, the soil began to dry up and blow away. Soil that had been more than 14 feet (4.3 m) deep was less than half that deep by the late 1940s. In some places in the Everglades Agricultural Area, only a couple feet of soil are left today.

Fire

It was fire in the 1940s, though, that really made people notice that something was wrong in the Everglades. With the saw grass and the land itself drying up, what had once been a river of grass became, in places, an endless field of fire. "Thousands, choking in acrid smoke, saw for the first time what the drainage of the Glades had brought to pass," wrote Marjory Stoneman Douglas.

Fire had always been an important part of the life cycle of the Everglades. In the past, it had been most frequently caused by lightning. Sometimes native people had set fires to control insects or to flush game from the Everglades. But with plenty of moisture in the ground, these fires had burned briefly and died out quickly. They had kept hardwoods out of the pine forests and kept trees from taking over the prairies.

But now the system was out of order. Besides lightning, fires were now started intentionally by sugarcane growers and accidentally by other people. There was no moisture in the ground, and the fires, once started, raged out of control. Fires burned for months at a time, smoldering in the

Sugarcane growers in the Everglades Agricultural Area set their fields on fire as part of the harvesting process.

dried-out muck. All plant life in the soil was destroyed, and countless animals were burned or starved to death.

Plants and Animals Suffer

Even away from the fires, the plants and animals of the Everglades were paying the price for human greed and folly. The ancient cypress trees of Big Cypress were cut down and hauled away for use in ship's hulls and bleachers. Mahogany was taken by the trainload to make furniture. Sable palms were destroyed for tasty "heart of palm" salads, and Spanish moss was gathered for stuffing mattresses and car seats.

Deer were slaughtered throughout southern Florida because farmers feared that tiny insects called ticks were carrying diseases from the deer to the precious cattle herds. Killing the deer didn't stop the infections, however, because nearly all the hairy wildlife of south Florida had ticks. If it weren't for the Miccosukee, who refused to allow the mass killings on their land, the white-tailed deer of the Everglades might have become creatures of the past.

Alligators were hunted for their skins. Turtles went into soup, and snail shells and butterfly wings were found more often in collections than in the wild. Animals that weren't killed outright suffered from loss of habitat and loss of food sources, as well as from the fires.

Then, in September 1947, another powerful hurricane hit southern Florida. Once again people shouted for more flood protection and a huge flood-

Not all the mahogany trees in the Everglades were cut down. This one is the largest mahogany tree still standing in the Everglades.

THE BOOK THAT OPENED PEOPLE'S EYES

A woman named Marjory Stoneman Douglas has been a resident of Florida since 1915. A lover of the Everglades, she worked for five years on a book called *The Everglades, River of Grass*. It was published in November 1947.

Douglas was the first writer to describe the Everglades as a "River of Grass." With those words she changed the way many people looked at the Florida wetland. In her book she explained that the entire Everglades was an interconnected ecosystem, and that when people interfered with one part, the whole could be affected. She described how the lives of the plants and animals and people of the region were related to each other. And most importantly, she described how all the life in the region depended on the fall of rain and the flow of water through the River of Grass.

The Everglades, River of Grass was an important part of the education effort on behalf of the Everglades. Slowly, more people came to understand that the Everglades was a unique place in which the flora and the fauna—the plants and the animals—rather than any spectacular mountains or waterfalls, were what made it worth saving.

control project was devised. It formally established the Everglades Agricultural Area and three water-conservation districts south of Lake Okeechobee. The plan called for the construction of numerous canals, pumps, and levees to direct water throughout the region. People would try to control what nature itself had once regulated.

This complex water-management project was designed to get water to the cities where it was needed, to keep excess water off the fields of farmers and ranchers, to replenish the aquifer with water stored in conservation areas, and, supposedly, to provide water for plants and wildlife in the Everglades. But the plants and wildlife didn't have the political clout that the other users had. Time would show that their needs were largely ignored.

Canal systems around the Everglades have diverted the natural water flow.

A National Park

As the drainage went on, a few individuals were becoming concerned about what all this was doing to the environment. As early as 1916, the Florida Women's Club had provided for the creation of Royal Palm State Park to protect a spectacular hardwood hammock southwest of Miami. Throughout the 1920s, landscape architect Ernest F. Coe called for the creation of a national park in the Everglades, but at first few people would listen. Many people still thought of the Everglades as a worthless swamp that would be of value only if drained and "put to good use."

Finally the wheels of politics began to turn, and

the long, slow process of creating a national park began. First, the U.S. government had to agree to accept the land if offered by the state of Florida. This legislation was passed by Congress in 1934 despite opponents who called it the "alligator and snake swamp bill."

The state of Florida passed a law enabling purchased land to be turned over to the federal government, but no money was provided to buy the land, much of which was in private hands. However, bit by bit, the land was purchased by the state of Florida and given to the federal government. On December 6, 1947, the Everglades National Park officially came into being, though the park's boundaries were not set for several more years.

In 1989, Everglades National Park was expanded by 107,000 acres (43,302 ha), providing further protection for critical water supplies.

Ernest Coe's original plan had called for some 2 million acres (809,371 ha) of marsh and swamp to be set aside. The park that was established was cut by a third to 1.4 million acres (566,560 ha) at the southern tip of Florida. This represented only about one-seventh of the total land of the Everglades. The park did not include the Kissimmee River, Lake Okeechobee, Big Cypress Swamp, or the northern stretch of the River of Grass. Coe was angry with this decision, realizing that without control of the waters that fed it, Everglades National Park would have many more battles to fight.

Drought and Flood

Where there had once been a 30-mile- (48-km-) wide channel for water to flow in sheets across the southern Everglades, there were now only a few places where floodgates along the Tamiami Trail let water into the park. And these gates were opened only "when water levels permit[ted]," meaning "when no one else wanted the water."

This led to a major crisis in the park. The protected portion of the Everglades became drier than ever before. Hundreds of animals were killed along the highway as they tried to reach the water in the conservation areas. Numerous plant and animal species became endangered.

The U.S. Army Corps of Engineers suggested that dikes be built along the southern edge of the park in Florida Bay to keep fresh water in the park

When the Army Corps of Engineers controlled the flow of water in central and south Florida, they prevented flooding. This caused drought in the Glades, and by the late 1980s, 90 percent of the wading bird population had disappeared.

from flowing into the bay. That would have meant the death of Florida Bay. Fortunately, they were told that the mangrove estuary that would be outside the dikes required fresh water, too, and no dikes were built.

Unnaturally high levels of water can be as damaging as levels that are too low. During the second half of the 1960s, there was heavy rainfall, and large amounts of water were released from the conservation areas into the park, creating flood conditions in areas where they would not normally occur. Alligator nests were destroyed. Numerous animals died, including over 5,000 deer.

To complicate matters even further, the timing of water delivery is just as important as the amount. The animals and plants of the Everglades require a periodic, regular rising and falling of water to survive. When floodgates were opened at the wrong times, birds would not nest, and many animals drowned.

In addition, the water coming through the floodgates from the conservation districts was increasingly polluted. That water came largely from Lake Okeechobee, which in turn had come from the Kissimmee River.

Lake Okeechobee was diked at its southern boundary to prevent flooding. After 1934, water no longer flowed freely southward toward the Everglades.

Ditching the Kissimmee

As part of the flood-control program around Lake Okeechobee, it was decided to try to "straighten out" the Kissimmee River. The U.S. Army Corps of Engineers began work on the project in 1961, completing it ten years later. Instead of a winding river with many "inconvenient" curves, known as oxbows, it became a straight channel. The 96-mile (154-km) meandering river became a 52-mile (84-km) canal, officially named C-38, unofficially called the Kissimmee Ditch. Farms and ranches sprang up along its sides.

Even before construction was complete, complaints were heard about the environmental damage caused by the project. The wetlands of the Kissimmee River Valley that had relied on the regular seasonal rising and falling of water levels were being destroyed. This meant the loss of habitat for many forms of wildlife, including fish, birds, reptiles, amphibians, and small mammals. Waterfowl stopped coming to the river. Bald eagles stopped nesting in the valley. Six species of freshwater fish disappeared entirely from the river.

The channelization also reduced the water quality in Lake Okeechobee. Water that ran from the farms and ranches into the canal was contaminated with fertilizers and manure. In the past, the long, winding river had often overflowed, and the wetlands around it helped absorb and filter out some of this. Now the water, with its heavy load of nutrients, dumped quickly into Lake Okeechobee.

In addition, water that had been used to irri-

The Kissimmee River winds lazily through the Florida countryside. The C-38 canal, shown above the river in the picture, diverted water from the river and made water travel more convenient but destroyed the wetlands.

When the Kissimmee River was "straightened," the water quality in Lake Okeechobee deteriorated. Runoff from farms was full of nutrients that cause algae to grow quickly, choking the lake.

gate the sugarcane fields south of Lake Okeechobee was also being pumped into the lake. That water, too, contained chemicals. All these chemicals caused algae bloom—or the excessive growth of algae. Because algae, like all plants, take in oxygen, there was little oxygen left in the water for fish, and many fish began to die. Dead fish not only hurt the fishing industry but meant less food for wildlife as well. The web of life in the Kissimmee River-Lake Okeechobee region was unraveling.

The Jetport that Almost Was

At the same time, along came the big idea that nearly spelled the end of the Everglades. It was a proposal by the Dade County Port Authority that a huge jetport be built in south Florida to complement Miami International Airport. The new airport would be bigger than several of the nation's biggest airports combined. It would be surrounded by an interstate highway, a high-speed rail system, and acres upon acres of homes and businesses. The site selected was on the southeastern edge of Big Cypress Swamp, just 6 miles north of the border of Everglades National Park.

When bulldozers and cement trucks rolled into the area, a public outcry arose. Environmental

BIRDS BEWARE!

Many of America's birds are threatened or endangered, and the birds of the Everglades loom large on the list. Poisons and other chemicals, water pollution, changes in water levels, and habitat loss are just some of the dangers faced by birds in south Florida.

In the 1950s and 1960s, pelicans (right) and bald eagles in the Everglades and in many other places were among the many birds found to be threatened with extinction by the chemical dichloro-diphenyl-trichloroethane, known as DDT. Farmers were spraying their fields with DDT because it was an effective way to kill insect pests. Unfortunately, DDT does not break down in the environment into harmless ingredients. Instead, it gets into water supplies and thus into fish. Birds that ate fish contaminated with DDT produced eggs with very thin shells that broke before the young hatched.

The use of DDT was banned in the United States in 1972. That has been a tremendous help to birds and other wildlife. However, in the Everglades, as in many areas, oil spills and other forms of water pollution continue to endanger wildlife.

Another problem for the many waterbirds of the Everglades is their dependence on regularly fluctuating water levels. For example, wood storks nest in late winter when water levels are low, and fish are concentrated and easy to catch. They require an enormous amount of fish to keep themselves and their nestlings fed during this time. Without just the right water levels, there are not enough fish, and the storks are unable to raise their young successfully. Human interference in the Everglades has caused dramatic changes in the water levels. There are many years when wood storks do not even build nests because they are aware that the water conditions are not right.

51

groups from across the country became involved. New studies were made, and the conclusions were clear: if this jetport were built, the Everglades would die. What wasn't ruined by drainage and development would be killed by pollution.

In 1970 the project was called off, but not before a single runway was built. Today, commercial and Coast Guard pilots sometimes use the runway for training. It stands alone on the former airport site as a reminder of how foolish humans can be.

Soon afterward, in 1974, Big Cypress Swamp became Big Cypress National Preserve so that it would be safe from such schemes in the future. But that action came too late to save an area along the western edge of Big Cypress. It presents a dramatic example of the environmental damage that can be caused by real estate development.

Where the stately cypress trees of Big Cypress had once stood, just to the west of the city of Naples, developers moved in and planned a huge subdivision to be called Golden Gate Estates. In order to offer homesites that would entice many buyers, 183 miles (295 km) of canals were dug and 807 miles (1,299 km) of roads were paved in the 1960s. Just as the canals of the Caloosahatchee and St. Lucie had lowered the water level of Lake Okeechobee, these canals bled off the waters of Big Cypress, lowering the water level by 2 feet (61 cm).

Even after all that work, only a small portion of the subdivision has ever been settled, because extensive sewage disposal was not provided, and many of the lots still flood during heavy rains.

Golden Gate Estates caused the death of many plants and animals without even keeping its promises to people.

Meanwhile, the estuaries of Florida Bay were suffering from a lack of fresh water, which was being diverted by canals along the west coast. The aquatic life of the bay requires a certain balance between fresh and salt water, and when that balance is severely disrupted, as it was then, the number of shrimp, oysters, crayfish, and other forms of marine life decreased. And when that happened, birds and other animals that depended on these creatures as food sources starved.

Brown pelicans *(right)* nest in the estuaries of Florida Bay *(below)*. About 300,000 birds come to the southern Everglades to nest each year.

Chapter Six

Saving the Everglades

It was clear that someone needed to speak for the Everglades when water decisions were being made in Florida. The South Florida Water Management District (SFWMD) was created in 1972 to operate the system of dikes, canals, and levees that control the water of south Florida. This agency now works with the national park and other groups to make the tough decisions on water availability and use throughout the region.

In an article in *Environment* magazine in 1984, SFWMD's director of Environmental Sciences, Walt Dineen, said, "Our biggest problem is that we are faced with balancing a diversity of interests, many in dire conflict with each other. We are consistently walking a management tightrope." These diverse interests include supplying fresh water to farms and cities, flood control in urban and rural areas, and restoring the natural ecosystem.

Today, while the SFWMD and the national park work hard to coordinate their efforts on behalf of the Everglades, Everglades National Park still gets only about half the fresh water it used to receive. Most of this comes directly from the rain that falls in the area each year.

Many experts agree that what the Everglades needs is a more natural water-delivery system. Studies are being made to determine how to match more closely the timing and amounts of water that natural sources would deliver to the Everglades.

Cleaning Up the Water

Since the late 1970s, the pollution damage to Lake Okeechobee has spread to the water conservation districts. Cattails replaced saw grass, and periphyton died and were replaced by an alga called microcoleus. These new plants tend to choke out the old ones, and they are not able to do the jobs the original plants did in the ecosystem.

Algae bloom caused by fertilizer and manure in Lake Okeechobee continues to be a very persistent

"At Duck Pond on the Park's Gulf Coast, it is still possible on summer evenings to see as many as 100,000 wading birds stream in to roost. . . . One need not be a bird-watcher to realize that Florida will be much poorer if such attractions are permitted to dwindle and vanish."

— William Robertson, Jr., in *Everglades—The Park Story*

The water of the Kissimmee River *(right)*, and other waterways in the Everglades, is full of nutrients from agricultural runoff. Now water hyacinths, which are not native to the area, and water lettuce clog the waterways.

problem. In 1986, more than 100 square miles (259 sq km) of the lake's surface was covered with green scum that suffocated snails, including the apple snails that are the Everglades kites' only food. Sports fish, such as perch and bass, died by the tens of thousands.

Scientists studying the problem have decided that part of the solution would be to replace the wetlands that have been dried up with artificial wetlands. As Florida Audubon Society president Dr. Bernie Yokel says, "These would function similarly to the native marshes in removing pollutants and moving water south into the water conservation areas leading to Everglades National Park." In other words, there is a need to keep wetland areas so that they can do what they have always done so well: act as natural filters to clean the water.

After being sued by the federal government for not enforcing pollution laws, the state of Florida is taking steps to fix the problem. Several areas of marshland were set aside in the Agricultural Area to filter fertilizers from irrigation water. Sugarcane growers were taxed to pay for the cleanup program. However, a large number of cane-growers have refused to pay the taxes and to abide by the new laws. Lawsuits and more delays will result—and this is only the first step in a long-term pollution-control program.

PLANTS OUT OF PLACE

Today, there are many kinds of plants found in the Everglades that literally don't belong there. Many of them were imported to grow in coastal cities and only accidentally found their way into the Everglades. Sometimes these plants, known as exotic, or introduced, plants, thrive so well that they begin to choke out native plants. Additional stresses on native plants, such as hurricane damage, sometimes allow these plants to take over. Then the whole ecosystem may be threatened. This process has been called a form of plant pollution.

For example, Australian pines were imported to line highways and canals half a century ago. Now, in the places where they grow, they block out the sun with their stems and branches and prevent other trees from growing.

Even more obnoxious is the melaleuca tree (left), also imported from Australia. This tree was planted in some places in south Florida because it soaks up a lot of water, but it also secretes a substance that drives away many animals—and it is extremely difficult to get rid of. Over 1 million acres (404,686 ha) of the Everglades are now considered infested with melaleuca.

Kudzu is an Asian vine recently found in one of the conservation areas. One of the world's fastest-growing plants, kudzu can grow as much as a foot per day in summer heat. Brought to a number of southern states in the 1930s to control soil erosion, kudzu quickly became a nuisance in many places. Biologists in the Everglades fear that such a plant could wreak havoc in a hardwood hammock if allowed to take root. Park rangers watch carefully for kudzu so that it can be eradicated immediately.

Restoring the Kissimmee

Slowly the official studies were made and reports were filed on the damage done by the creation of the Kissimmee Ditch. A recommendation to restore the river and its valley to their original state was made. While the cost of channelization had been just over $30 million, the cost of restoration is estimated to be $350 million. Beside the cost of actually dechannelizing the river, there are now the added costs of buying back land that was developed when the floodplain was drained.

In August 1983, Governor Bob Graham introduced the Save Our Everglades program. Graham stated that the purpose of the program was to "provide that, by the year 2000, the Everglades will be more like it was in 1900 than it is today." A large part of this program hinges on restoration of the Kissimmee River.

A demonstration project was conducted from 1984 to 1985. Part of the canal was blocked so that the river was forced back into a stretch of its natural channel. Scientists noted an immediate improvement in wildlife conditions.

The South Florida Water Management District, the U.S. Army Corps of Engineers, and many agencies continue to work on this mammoth project trying to undo the damage done in channelizing the Kissimmee River. It's going to take a very long time to make things right.

As part of the Kissimmee River restoration project, begun in mid-1984, three steel dams, called weirs, were installed to divert water from the canal back into the original winding river channel and floodplain. The gap in the center allows small boats to continue to use the main C-38 channel.

Friends of the Everglades

One of the most promising signs for the Everglades is the growing number of people who are concerned about saving it, such as Marjory Stoneman Douglas, known as "The Grandmother of the Glades." During the controversy over the jetport, Marjory Douglas, who was then 78 years old, created a group called Friends of the Everglades. For $1, anyone could join and show their support for the plants and animals of the tropical wilderness.

Some of the group's most active members decided to organize protests against development in the Everglades. Led by Marjory herself, they became known as "Marjory's Army." Their defeat of the jetport project was a major victory.

Since that time, the Friends of the Everglades have been busy on many fronts. They make sure that the state and federal governments are acting in the best interests of the Everglades. In 1991 the governor of Florida came to Marjory's home to sign a bill requiring a cleanup of the Everglades. Education programs are run at the Marjory Stoneman Douglas Nature Center in Key Biscayne.

The Seminole and Miccosukee also wish to have the Everglades preserved. Many earn a living in various aspects of the tourist industry—from alligator wrestling to offering airboat rides and selling crafts. These activities depend on a lush and thriving Everglades for people to visit and enjoy.

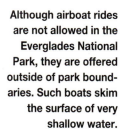

Although airboat rides are not allowed in the Everglades National Park, they are offered outside of park boundaries. Such boats skim the surface of very shallow water.

There are many others who care about the Everglades. The Florida Audubon Society, the Florida Wildlife Society, and the Everglades Coalition are just some of the organizations that keep an eye on the situation in the River of Grass and its surroundings. They give voice to the plight of the plants and animals who cannot speak for themselves.

Manatees are safe from humans and their boats in protected areas such as Homosassa Springs State Wildlife Park.

Endangered Animals

Approximately 100 manatees die each year in the waters along the Florida coast. Many of these deaths are caused by injuries from boats. Rare is the manatee that doesn't have huge scars from boat propeller cuts. Some manatees die from venturing into waters that turn too cold in winter. Others die from eating polluted vegetation or from becoming entangled in fishing lines and garbage dumped in the water.

In 1978 the entire state of Florida was named a manatee sanctuary. In special manatee protection zones, boats must travel at slower speeds when manatees are present.

The American crocodile depends on the health of Florida Bay, and Florida Bay has big problems today. For example, a particular kind of sea grass has been dying in large areas due to a decrease in the flow of fresh water into the bay. This sea grass was the basic food supply for many small organisms that were, in turn, the food supply for larger sea creatures that are part of the food chain for crocodiles. This is a good example of how complex an ecosystem can be, and how problems for one area of an ecosystem can create severe difficulties for other parts of the ecosystem as well.

The loggerhead turtle has to deal with condominiums, people, and vehicles on the beaches it has traditionally used to nest. Lights on the beach can distract the newly hatched turtles into heading toward land instead of the sea after they hatch. When this happens, the turtles die from heat and lack of moisture—if raccoons don't get them first.

Loggerhead turtles are now a protected species. When the young turtles hatch *(left)*, they usually head straight for the water.

And in many of their nesting areas, the population of raccoons has risen due to the elimination of many of their natural predators.

Since the early 1990s, efforts have been made to prevent the drowning of sea turtles in shrimp nets. All sea turtles in American waters are now protected by the Endangered Species Act.

Many of the Florida panthers' problems can be traced to an increase in human population. The panthers' problems include death on the highway and death by mercury poisoning. A captive-breeding program and reintroduction of panthers to suitable breeding habitats are two of the ways in which people are trying to help panthers. But sometimes panthers are hurt in the process of being helped. Some people think that the collaring program, which scientists use to monitor a panther's habits, is inherently cruel because panthers are cats, and cats hate collars.

From the early 1900s until World War II, the number of wading birds in Florida increased to almost their original numbers. Since the 1940s, however, the diking of Lake Okeechobee and draining of the Everglades have caused many bird populations to drop to about 10 percent of their original numbers.

The major problem for waterbirds in south Florida is the loss of feeding areas. Even if they find safe places to nest in places like Everglades National Park and Corkscrew Swamp Sanctuary, these habitats may not provide the quantity of food the birds need during nesting season.

The Only Everglades

The Everglades certainly has changed greatly in the years of human involvement. It has been drained, dredged, dried, tilled, developed, ignored, and then overmanaged. None of these human activities has helped this spectacular spot.

Everglades photographer Joel McEachern remarked in a 1991 article in *Environmental Action* magazine, "Many places I photographed just five years ago are gone. Development and neglect have taken their toll across the state, particularly along the coast where wetlands once thrived."

The Everglades has been in danger of dying altogether, and it has been repeatedly saved. It is still in danger. In recognition of its unique status, the United Nations has named the Everglades both an International Biosphere Reserve and a World Heritage Site. Many people are now devoting their lives to protecting the Everglades. As Marjory Stoneman Douglas said, ". . . these are the only Everglades in the world."

Visitors to Everglades National Park enjoy watching and photographing birds at Mrazek Pond.

GLOSSARY

algae bloom – the excessive growth of primitive, usually one-celled plants from the addition of fertilizers and other nutrients to the water. This depletes the water of oxygen, killing animal life.

aquifer – an underground rock formation in which all the spaces are filled with water.

detritus – loose matter, such as rock particles or small bits of plant and animal material, resulting from erosion, disintegration, or destruction.

dike – an embankment raised to prevent a river from flooding; also called a levee.

ecosystem – a plant and animal community together with their physical environment, considered as a unit.

epiphyte – a rootless plant that uses another plant for support; an air plant.

estuary – the area where the fresh water of a river mingles with the salt water of the ocean.

habitat – the area or environment in which a living thing is normally found.

hammock – an island of land in a marsh, such as the Everglades, which is largely covered with hardwood trees.

head – a round hammock on which one specific kind of tree is found.

larvae – the caterpillars and other creatures that are an early stage in the development of many insects, corals, and other invertebrates.

levee – *see* dike.

peat – soillike substance formed by layers of once-living matter.

periphyton – small algaelike plants that grow in masses in wetlands.

slough – a shallow depression, or low area, in which water concentrates.

strand – a long, narrow stand of trees.

wetland – a lowland area that is saturated some or most of the year with water. Marshes, swamps, and bogs are types of wetlands.

FOR MORE INFORMATION

Books

Carr, Archie. *The Everglades*. New York: Time-Life, 1973.

Caulfield, Patricia. *Everglades: Selections from the Writings of Peter Mattiessen*. New York: Sierra Club/Ballantine, 1970.

Challand, Helen B. *Disappearing Wetlands*. Chicago: Childrens Press, 1992.

Cowing, Sheila. *Our Wild Wetlands*. New York: Simon & Schuster, 1980.

Cox, Eugene. *Everglades in Pictures; The Continuing Story*. Las Vegas, Nev.: KC Publications, 1989.

Derr, Mark. *Some Kind of Paradise; A Chronicle of Man and the Land in Florida*. New York: William Morrow, 1989.

Douglas, Marjory Stoneman. *The Everglades: River of Grass*, revised edition. Sarasota, Fla.: Pineapple Press, 1988.

Garbarino, Merwyn S. *The Seminole*. New York: Chelsea House, 1989.

George, Jean Craighead. *Everglades Wildguide*. Washington, D.C.: U.S. Government Printing Office, 1972.

Golia, Jack de. *Everglades; The Story Behind the Scenery*. Las Vegas, Nev.: KC Publications, 1978.

Harris, Bill. *The Everglades: A Timeless Wilderness*. New York: Crescent Books, 1985.

Robertson, William B. Jr. *Everglades: The Park Story*. Coral Gables, Fla.: University of Miami, 1959.

Sosin, Mark. *The Everglades*. New York: W.H. Smith, 1991.

Stone, Lynn M. *Florida*. Chicago: Childrens Press, 1987.

Videos

Audubon Video. *Wood Stork: Barometer of the Everglades*. Vestron Video.

Nature Scene. *The Everglades and Sanibel Island*. PPI Entertainment Group.

INDEX